Kubutambahan
Sengsit
Tejakula
Singaraja
Penulisan
Kintamani
Lake
Batur
Lake
Bratan
Terunyan
Bedugul
Penelokan
Baturiti
Gunung Agung
Seribatu
Besakih
Kendang
Penebel
Karangasem
Bangli
Tenganan
Sangeh
Iseh
Manggis
Ubud
Peliatan
Bedulu
Marga
Teges
Mas
Padangbai
Gianyar
Kapal
Gelgel
Kusamba
Kediri
Lukluk
Batuan
Sempidi
Batubulan
DENPASAR
Sanur
Karengsari
Kuta
NUSA
PENIDA
Airport
Benoa
BUKIT
BADUNG
Nusa Dua

Fruits of Bali

Peint d'après nature par Mme Berthe Hoola van Nooten à Batavia.

Chromolith. par G. Severeyns Lith. de l'Acad. Roy. de Belg

ZALACCA EDULIS. REINW.

Fruits of Bali

FRED AND MARGARET EISEMAN

Published by Periplus Editions (HK) Ltd.
Distributors:
Singapore & Malaysia : Periplus (S) Pte Ltd, Farrer Road
P.O. Box 115, Singapore 9128
Indonesia : C.V. Java Books, Cempaka Putih Permai
Blok C-26, Jakarta Pusat 10510
Benelux : Nilsson & Lamm B.V., Pampuslaan
212-214, 1382 JS Weesp, The Netherlands

PUBLISHER: Eric M. Oey
EDITED BY: David Pickell
COVER: Pete Ivey
DESIGN: Pete Ivey and David Pickell
PRODUCTION: David Pickell
PHOTOGRAPHY: Fred and Margaret Eiseman

Library of Congress Catalog Card Number: 88-61078
ISBN: 0-945971-02-8

First printing 1988
Second printing 1990
Third printing 1992
Fourth printing 1994

Printed In Singapore

Frontispiece: Bali's most "typical" fruit, salak, in a 19th Century print by G. Severegne after a drawing by Berthe Hoola van Nooten. Courtesy of Antiques of the Orient, Singapore.

Dedication

We dedicate this book to Njoman Oka, our friend since 1961, with appreciation for all the things he has taught us.

The first Balinese we met and our first guide, Njoman even then had the reputation of being the premier guide in Bali, having helped anthropologists whose names are now legendary. In those days guiding was more arduous than simply taking a bus load of tourists to the barong show at Batubulan and then to lunch at a restaurant in Kintamani. Guiding meant exploration via byroads not well-worn by tour groups; it meant getting up at midnight to attend an unusual ceremony—essentially, it meant a hands-on introduction to a culture we never dreamed existed.

Njoman had a wealth of experience upon which to draw. He had been, among other things, a resistance fighter against the Dutch, Bupati of Tabanan, head of a school and director of a successful travel agency.

Njoman's reputation has been enhanced in the 27 years since our first encounter. He has been our friend at countless events and an advisor on every imaginable subject. It is no exaggeration to say that our interest in Bali was originally sparked by that first meeting. And this spark kindled a flame that has burned bright in our minds ever since.

Njoman has taught German at his beloved Saraswati school and remains active to this day guiding selected clients. Many members of his large family are reputable guides. He goes several times a week to Tabanan where he actively sponsors a Pasantian, a group of villagers who meet regularly to perfect their knowledge of, and ability to read and chant the sacred Hindu stories and scriptures. He writes frequent letters to the editor of the Bali Post, taking well-aimed jabs at developments he feels are detrimental to Bali's long term cultural preservation. He, of all the Balinese we know, has most successfully straddled two cultures, taking the best from East and West.

To Njoman Oka, we say:

Thank you for introducing Bali to us, and for your continued help and valued counsel. Had it not been for you, we probably would never have had the experiences that have occupied our minds and our pens for so many years.

Contents

Introduction

Shopping for some of Bali's interesting native fruits is just as necessary to complete your experience here as seeing the Batubulan Barong Show or visiting Kintamani, the Monkey Forest and Tanah Lot. And what's more, it takes less time, and is a lot cheaper. If you haven't yet tested the defenses of a Durian or tried to figure out how to peel a mango, you haven't yet had the full "Bali experience."

The fruits of Bali are not unique to the island. You will find the same varieties throughout much of tropical and sub-tropical Asia. But because Bali is so small, during the course of a year you can find practically all of them, freshly picked, waiting for your inspection or purchase at a local village fruit market within a few hundred meters' walk or a few hundred rupiahs' bemo ride from wherever you are staying. And what you will be eating is not imported; almost all of the fruit sold in Bali is homegrown.

Inspecting Balinese fruits does not consist of ordering them already peeled and cut up and nestled in a bed of shaved ice on the best crystal at a fancy hotel. Inspection means a one-on-one, eyeball-to-eyeball, closeup encounter with fruits piled on the counter of a stall in the market or spread out on a mat on the edge of a street on market day. This is the way the Balinese buy them.

Buying fruits is much cheaper than buying paintings or woodcarvings, and you will have just as much fun bargaining with the sellers—probably the only Balinese with whom you will interact who have not been trained to cater to the peculiarities of foreigners. Body language is quite sufficient. And you are not obligated to buy anything, although you are missing something if you don't.

We have included in this book all of the fruits that a visitor is likely to encounter in Bali—unless he makes a deliberate effort to search for the most unusual ones. We have omitted some fruits commonly found in the West, such as

Left:
Fruit stalls at the market in Sukawati.

grapes and strawberries. These are grown in Bali, but chances are that they are available at home and that you are already familiar with them. We have also omitted dozens of scattered local fruit varieties because these almost never find their way to market. As soon as they are ripe little boys knock them off the trees with stones and eat them on the spot. These little boys are true fruit lovers.

In Bali, in almost all cases, fruits are grown by small landholders as a source of income incidental to some other principal occupation. It costs nothing to plant a few fruit trees in the houseyard. The plants have not been specially bred. They have not been selected on the basis of how well the fruits that they yield will withstand shipping or how pretty they are, as is the case in many Western markets. They are not sprayed, fertilized, cultivated, and nursed along until the fruits are just underripe and of perfect shape, so that they will reach the market at the peak of quality. And so you will see that Balinese fruits will often not have the unblemished exteriors that seem to be necessary to attract Western buyers. The lack of careful genetic selection also results in considerable variability in taste and quality.

Most fruits are raised on scattered small holdings, often quite far from the main roads. With one or two exceptions there are no large fruit farms or orchards such as one finds in the West. Fruits are brought to market from the farms by professional collectors who have regular territories and customers. So, even if one were able to find an area where a particular fruit is grown in abundance, and even if he arrived at the peak of the season, he would likely find no fruits on the plants because of their recent harvest by the collectors.

The greatest abundance and variety of fruits is not at their point of origin, but rather, at the markets where they are sold. If there is any fruit at all at the source, it is likely to be more expensive there than at the markets because of its scarcity and because the farmer never retails small quantities of his produce.

The exceptions to this general rule are the extensive vineyards in the area to the west of Singaraja and the small citrus orchards in the Kintamani area, both of which are readily accessible and both of which have concentrated areas of planting.

Collectors and farmers usually harvest their crop together whenever the collector happens to arrive. The farmer is paid in cash, and the collector hauls off the fruit in his truck—often to other collection points, and then to market. Almost all markets in Bali have a special day when most people come to shop and sell. Market day is called *pekenan* in Balinese, and it usually occurs every three days at a given market. The three days of the market week, called the *tri-wara*, are *Pasah*, *Beteng*, and *Kajeng*. Sometimes *Beteng*, the second day of the three-day week, is also called *pekenan*. A list of the principal market days in various towns is given in the sidebar on the following page.

Shopping is best on *Pekenan*. Some smaller village markets are only open on that day, so check before you go. The larger markets, such as the huge *Peken Badung*, the central market in Denpasar, are open and active every morning. There usually is not much going on in any market in the late morning and afternoon.

Collectors sell directly to their regular customers at the village markets. They arrive very early in the morning, well before dawn, and this is the best time to go shopping. The very best fruits are quickly bought up and are likely to be gone by 7:00 am. The wholesale mark-up averages about 20 percent, but this depends somewhat upon the availability of the fruit, the demand for it, and whether or not there are any important forthcoming Hindu-Balinese religious holidays, for example, Galungan, at which times prices jump by 20 percent.

Almost all fruits are seasonal. The major exceptions are coconuts, papaya and bananas. The lengths of the seasons for various fruits are quite variable. Some, such as jackfruit, are at their peak for about three months, and there is seldom a time when one cannot find at least some in a market. Others, such as *cereme*, have a season of only about two weeks or so, and it is almost impossible to find them in any market at any other time.

The Balinese normally think and speak of lunar months, *sasih*, rather than the calendar months that are familiar to Westerners. The reason for this is partly tradition and partly religious, since some Balinese religious festivals are geared to the cycle of lunar months. The lunar year begins in March on the day after the new moon. This is the start of lunar month number 10. The next month is number 11, then 12, and then the cycle begins again, usually in July, with month number 1 and so on. The names of these lunar months are the Sanskrit words from 1 to 12. The rainy season, from about November to March, occurs during months 11-3. In general, fruits are most abundant, larger, better, and cheaper during the rainy season.

PLACE	MARKET DAY
Badung(Denpasar)	*Beteng*
Bajra	*Kajeng*
Bangli	*Pasah*
Baturiti	*Kajeng*
Candi Kuning	No special day
Gianyar	*Beteng*
Kapal	*Kajeng*
Kediri	*Kajeng*
Klungkung	*Pasah*
Kuta	No special day
Luwus	*Kajeng*
Payangan	*Kajeng*
Penelokan	*Kajeng*
Sukawati	*Kajeng*
Tegallalang	*Beteng*
Ubud	*Pasah*

Because of the variations in soil and climate, a given fruit may have a slightly different peak season in different areas. The effect is to lengthen the time during which the fruit is readily available at the markets, since all markets are serviced by collectors who range over wide areas of Bali.

Market sellers normally rely upon a retail mark-up of 20 percent. However, there is no such thing as a fixed price in Bali—for anything—and bargaining is an absolute must. Balinese women know very well the normal prices for most fruits, and the sellers know that they know. The sellers also know that visitors to Bali do not know. They also know that the tourist is not likely to buy very much, if anything. And so the visitor can count upon the asking prices being about double the usual. But, sometimes the tourist can bargain the price down below that which a Balinese would pay, since the latter would not wish to offend by aggressive behavior. The ground rules for bargaining for fruits are the same as for anything else. A knowlege of Balinese or Indonesian helps but is not necessary. Body language is universal, and fingers are good indicators for amounts of money.

In the larger village markets most small fruits are sold by the kilogram. Those that grow on bunches, such as bananas, are sold by the bunch. Larger fruits are sold by the individual piece. Roadside *warungs* and small markets usually sell most fruits by the piece rather than by weight. For those not accustomed to the metric system, one kilogram is about 2.2 pounds. One pound is about 450 grams. Prices are usually quoted in rupiahs per kilogram. With the value of the rupiah at about 1,660 per U.S. dollar, a price of Rp 1,000 per kilogram is equivalent to about US$ .25 per pound.

By Western standards, fruits in Bali are cheap. About the most you could possibly pay for any sort of fruit, even if being charged outrageously high tourist prices, is about Rp 2,000 per kilogram. Most prices are well below this figure. You can buy a delicious bundle of all that you and your family could possibly eat in a day for the cost of a single dessert at a fancy restaurant. And, since fruits come with peels, you need not worry about hygiene.

Fruits that grow on bunches, like bananas, are usually sold in bunches. If you want to buy an individual banana, the seller will have to cut it off the bunch, and you can expect to pay more. As a rule of thumb, the price of a single banana to a Balinese is usually about Rp 50 (about 3 US cents), and a single coconut costs about Rp 200 (about 12 US cents). Figure on being overcharged by about 50 percent unless you are persistent. Fruits are cheaper per unit if you buy more than one.

Fruits are picked on a schedule that is determined by the convenience to the collector, not by careful calculation so that the peak of ripeness will occur upon market day. With some fruits ripeness is obvious by inspection, as is the case with bananas and grapes. With most others it is quite difficult to determine whether a fruit is ready to eat or not. Therefore you must pay careful attention if you want to make sure of buying fruits that are sweet and ripe. In the case of small fruits, such as citrus, sapodilla, and water apple the easiest way to determine ripeness is by buying just one and tasting it. For larger and more expensive fruits, the seller usually has a sample cut open and will let you taste it for free. As a visitor you can, of course, expect the seller to try to get rid of her inferior fruits on one who has had little experience.

Prices vary considerably during the season, depending upon availability. Prices are always higher on the more important religious holidays. They are highest at Galungan, somewhat higher at Kuningan, and a bit higher than normal on some of the Tumpeks. Prices are likely to be lower in the afternoon, when the supply of customers is waning, and they are always higher at *warungs* than at the village market—even for Balinese—because the people running the *warung* buy the fruit at the market too. And prices border on the ridiculous at all regular tourist bus stops—for obvious reasons. The same is true about prices at supermarkets. Prices are always higher the further the market is from the sources of the fruit, because of transportation costs. Mark-ups are always higher for fruits that are likely to spoil readily, for example soursop, or bananas, than they are for those that have good keeping qualities, like coconuts. There is a smaller variation in seasonal prices for those fruits, such as Caffier lime and pomegranate, that are normally scarce even at the peak of the season. The authors assume no responsibility for the reader being overcharged when buying a banana.

—Fred and Margaret Eiseman

Opposite: The celebrated durian, *most notorious of all Southeast Asian fruits. (Print courtesy of Antiques of the Orient, Singapore.)*

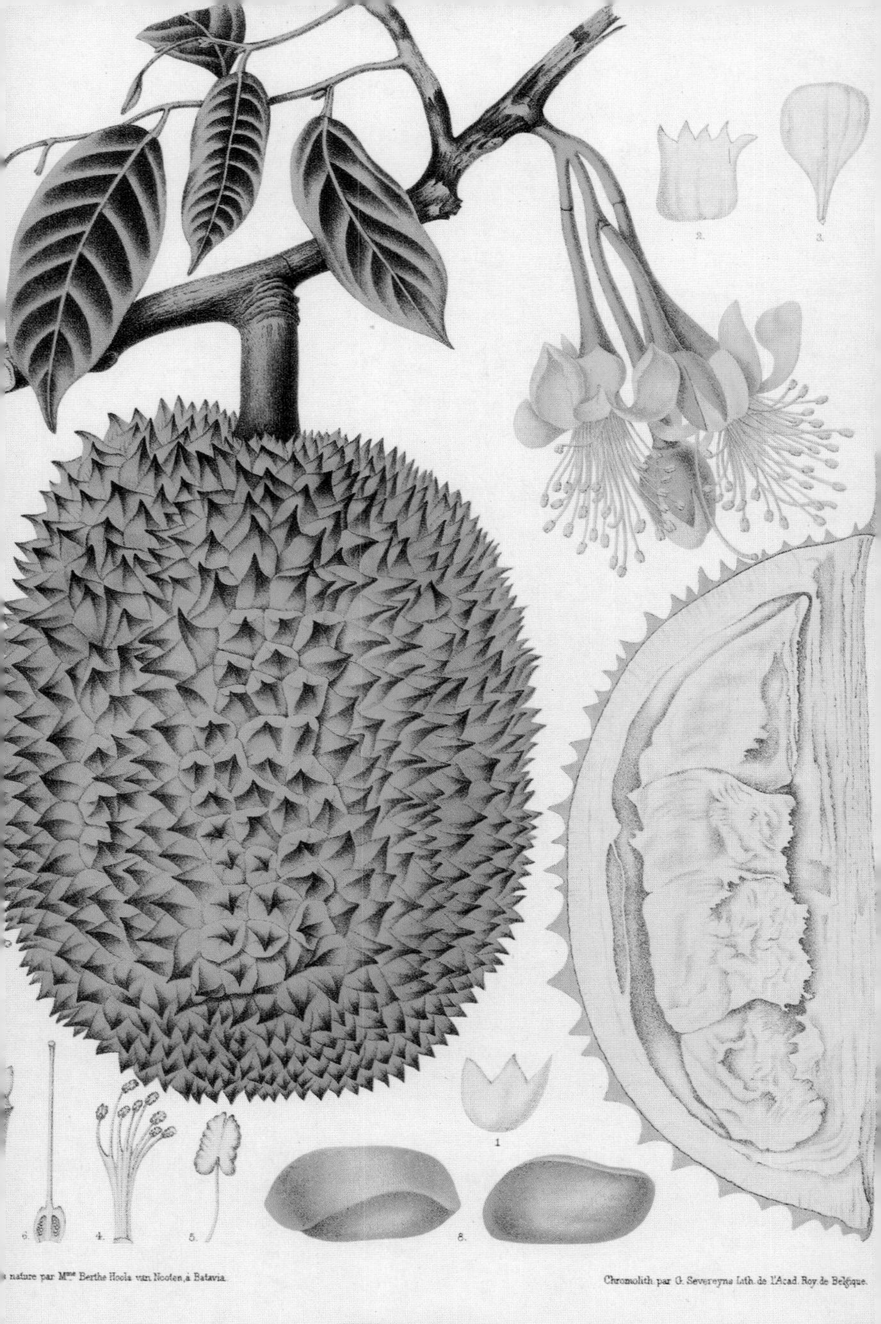

DURIO ZIBETHINUS. L.

Apple

The apple is of course a temperate zone fruit; one cannot expect to find world class apples in the tropics, and one does not get them. But sometimes a so-so apple is better than none at all. In fact, the local variety is not all that bad—juicy and crisp if rather tart, and quite refreshing when juiced. In sweetness, however, it can never rival the taste of a Red Delicious.

Practically all of the apples you see in Bali are actually grown in the hilly Malang area of East Java. There are a few orchards in the Besakih area, but they contribute only an insignificant percentage of the total supply.

The apples here are all of one variety, having a green skin with red blotches and streaks. The flesh is almost pure white and the size is small to medium, 7 to 9 cm across.

BALINESE: *apel*
INDONESIAN: *apel*
LATIN: *Malus communis*; Family: *Rosaceae*

Avocado

Also butter fruit, alligator pear

The avocado was introduced to the West by the Spanish conquerors of Mexico, and the name is a corruption of the Aztec word *ahuacatl*, meaning both fruit of the avocado tree and testicle. As far as can be determined, the tree was introduced to Indonesia in the middle of the 18th century.

BALINESE: *apokat, adpokat*
INDONESIAN: apokat
LATIN: *Persea americana;* Family: *Lauraceae*

The Balinese varieties are not very different from those found in the West. They grow best at medium elevations, such as on the slopes of Mount Batur in the vicinity of Bedugul. They vary in size, but are usually pear-shaped with a single, walnut-sized seed embedded in the yellow, buttery flesh.

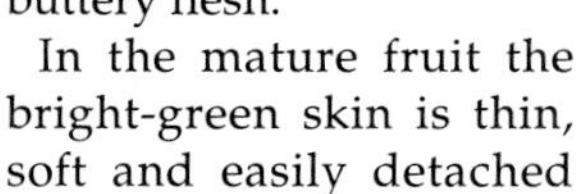

In the mature fruit the bright-green skin is thin, soft and easily detached from the flesh. The flesh has a faint odor and contains a high percentage of an oil similar to olive oil. The fruit is rich in vitamin B and contains more than three times the protein of apples or pears.

The Balinese most commonly eat the avocado along with other fruits, cut up and mixed with chipped ice and a brilliant red sugar syrup. This concotion is called *es campur*, which simply means "mixed ice" and it can be purchased from push carts, *warungs*, and restaurants all over Bali. Mixed with sugar and water, avocados also make a common iced drink, known as *es apokat*.

In hotels catering to foreigners the avocado is of course served in more familiar ways—sliced in a fresh salad or stuffed with succulent shrimp and crab.

Banana

Bananas have been common throughout the temperate countries of the West ever since the advent of refrigerated shipping in the 1920s. Today most of the bananas available in these countries are "plantation bananas" from Latin America or the Philippines. Most Asian bananas fall into the category of "backyard bananas" grown for local use.

BALINESE: *biu*
INDONESIAN: *pisang*
LATIN: *Musa* sp.; Family: *Musaceae*

The plantation banana is a variety that offers a high yield and a fruit that travels well. But these qualities do not necessarily result in the best tasting banana. In Bali, bananas come in dozens of different varieties, offering a wide range of sizes, textures and tastes.

The delicious *biu susu* or "milk banana" is no bigger than a man's thumb. The "king banana" or *biu raja* is just slightly smaller than the common supermarket variety. The peel of the *biu udang*, or "shrimp banana," is red. That of the *biu gading*, or "ivory banana," remains green even when the fruit is ripe. The "papaya banana," *biu gedang saba*, is fat like a small papaya, while the *biu kayu* or "wood banana" is quite thin. And the *biu batu* or "stone banana" is a real surprise—it is full of hard seeds!

The banana plant, actually a giant herb, is put to good use by the Balinese. The soft "trunk," really a stem, is used as animal fodder. The stems are also used in the *wayang kulit* puppet play to hold the puppets. The leaves serve as plates or all-purpose wrappings for food and offerings.

Breadfruit, Breadnut

Neither breadfruit nor breadnut conforms to our usual idea of a fruit since both must be cooked before eating. Although the trees are botanically very different, they are almost identical in appearance—tall, with large, spreading, multi-lobed and pointed leaves. The fruits are also rather similar-looking, both ellipsoidal in shape and dark green in color, but here the similarities end.

The skin of the breadnut is covered with stout, black fingers. The breadfruit has smooth skin but seems rough because it is covered with circular splotches of brown-black protrusions. The breadnut is full of large seeds; the breadfruit is seedless. The interior of both is tannish-white and spongy, but not soft.

There isn't much one can say about the gastronomical qualities of these fruits. They are good sources of staple starch, but are devoid of any distinctive flavor.

BALINESE: *sukun* and *timbul*
INDONESIAN: *sukun* and *timbul*
LATIN: *Artocarpus communis*; Family: *Moraceae*

Caffier Lime

Also ichang lime

This fruit is used mainly for its medicinal value and in cooking, and you will only find it in the spice section of the market. It is scarce and rather expensive compared to the other common citrus varieties.

The caffier lime is the most peculiar-looking of all Balinese citrus and is distinguished by its deeply and irregularly wrinkled skin and pear-like shape. The skin is green with a pronounced nipple at one end. The flesh is pale green with small seeds and has a sour to bitter taste.

The fruit is an ingredient in a number of traditional Balinese medicines. The juice is also used to wash hair.

BALINESE: *juuk purut*
INDONESIAN: *jeruk purut*
LATIN: *Citrus hystrix*; Family: *Rutaceae*

Cashew

The cashew tree is native to tropical America and was apparently brought to Asia by the Portuguese. The fruit is soft and edible, if rather tart, though it is the nut produced at the end of the fruit that is prized around the world for its delicious taste and high oil content.

A hard shell encapsulates the nut, and around it there is also a thin tissue containing an acrid, oily sap that is an irritant and diuretic, and is used in many parts of the world for its medicinal properties. In Brazil, the juice of the fruit is used to make a fermented drink. In Java and Bali, the young leaves of the tree are eaten raw with rice.

BALINESE: *nyambu menté*
INDONESIAN: *jambu meté*
LATIN: *Anacardium occidentale*; Family: *Anacardiaceae: Moraceae*

Coconut

One can almost say that Balinese civilization is founded upon the cultivation of four plants: bamboo, bananas, rice, and coconuts.

The Balinese would consider the dry and aged coconut that reaches a western supermarket to be fit only for feeding to animals or pressing for oil. The Balinese distinguish between young coconuts and the mature coconuts familiar to us in the West.

The Balinese call the former *kuwud* and a coconut reaches that stage after about 9 to 10 months. The husk has not har-

BALINESE: *nyuh*
INDONESIAN: *kelapa*
LATIN: *Cocos nucifera*; Family: *Palmae*

dened into the dry, fibrous mass that cushions the 30 meter fall of the nut to earth but is still moist and can be easily cut off with a sharp, curved knife. The interior is full of delicious coconut water and the white meat is thin, soft and moist—not hard or dry. It can be easily scooped out of the inside of the shell with a small chunk of coconut shell or a spoon and eaten after the juice is drunk. The sweetened water of the young coconut, mixed with shredded soft coconut meat and ice, is sold everywhere in Indonesia as a refreshing drink called *es kelapa muda.*

The white, hard, flesh of the mature nut—copra—can be dried and stored without spoiling. If cut up and boiled in water, the coconut oil—the standard cooking oil of Asia—rises to the surface and can be skimmed off. If this white meat is shredded and mixed with a little hot water then squeezed, coconut cream is the result. Many curried dishes use this as the principal ingredient, and it is also used to make a number of delicious desserts. Like dairy cream, its cholesterol content is extremely high.

A type of palm sugar is made by boiling the juice of the incised coconut flower, and if allowed to ferment naturally, the juice turns to a mildly alcoholic drink called *tuak.*

Immature coconut leaves, still pale yellow because they have not yet been exposed to sunlight, are an essential part of virtually all Balinese offerings. Older leaves are woven into mats. The shell of the nut is used as a drinking vessel and also makes the best charcoal for grilling skewers of *sate*. The fibrous husk of the mature coconut can also be made into mats and cordage and is widely used in Bali as a fuel. The woody spathes of the flower are also gathered as firewood. The wood of the trunk is a standard building material for the rafters and beams of a house. The stump is often carved into a durable and extremely heavy chair. The bud at the top of the tree is eaten as a vegetable.

Cucumber

Cucumbers are a stock ingredient in Balinese kitchens and *warungs* and the Balinese distinguish two varieties: *ketimun gantung* or 'hanging cucumber'—the common cucumber of the Western world—and the giant *ketimun guling*.

BALINESE: *ketimun gantung, ketimun guling*
INDONESIAN: *timun, ketimun*
LATIN: *Cucumis sativus* L. Family: *Cucurbitaceae*

The Balinese eat cucumbers much more frequently than Westerners. They are served raw as an accompaniment to rice and the usual side dishes. Sometimes they are cooked as a vegetable. When still raw they are used to make *rujak* (see page 55). Pickles are made by soaking slices of cucumber in a vinegar that is made from palm wine.

Whenever there is a special event—a cockfight, an *odalan* temple ceremony, or a football game—you can be sure that someone will be selling *ketimun gantung* and that lots of people will be eating them plain, the way one might eat an apple. Most Balinese like to bite off one end, eat the inside of the fruit, and spit out the bitter skin.

Ketimun guling, the giant cucumber, is usually the size of a small watermelon. The basic color is green, sometimes very dark, and is striped with yellow or orange. The fruit is often eaten sliced in section like a melon. Otherwise it is peeled, sliced into cubes, and eaten with white sugar. It is not normally eaten with rice, like *ketimun gantung*. This fruit is common, inexpensive, and available in every market and in roadside stalls.

Custard Apple

Also sweetsop, sugar apple

The custard apple has an excellent flavor but is not as popular among the Balinese as its close relative, the soursop. The Balinese name is a corruption of the original Malay name *srikaya* meaning "rich in grace."

The small trees that are native to the West Indies grow quite well on the dry Bukit region of southern Bali, where the fruit becomes ripe beginning in September.

At first glance the fruit looks something like an artichoke. It is green in color and is covered with rough pentagonal protrusions. The skin of the raw fruit is hard but it softens greatly as it ripens. The interior has a small conical core, but the rest is completely filled with white pods containing large seeds. The flesh is creamy and has a texture something like that of soft cheese, but is a bit fibrous. The taste is sweet and delicate, but since a fair percentage of the interior is seeds, you get little fruit for much work.

BALINESE: *silik, silikaya, sekaya*
INDONESIAN: *srikaya*
LATIN: *Annona squamosa*; Family: *Annonaceae*

Durian

Durian, from the Malay *duri* meaning "thorn," is the most controversial of all tropical Asian fruits. Some prize it as the equal of nectar from the gods, while others agree with Sir James Scott, who in 1882 wrote that "the flavor and odor of the fruit may be realized by eating a 'garlic custard' over a London sewer." Most Indonesians love the durian with a passion, and for that reason it is one of the most expensive of fruits and a great delicacy. The strong smell, however, has a tendency to put Westerners off, and the fruit is generally banned from all hotels and restaurants catering to them as a result.

A native of mainland Southeast Asia and Indonesia, durians grow best at medium elevations in warm, humid areas. During the season, which begins in late September and lasts through December (the latter part of the dry season), durians are plentiful in the markets and roadside stalls. At other times of the year, they can also sometimes be found though they are even more costly than usual and not as flavorful.

The average durian is about the size of a coconut, though the size can actually vary greatly. The extremely hard, fibrous rind is covered with many sharp spikes. The interior is divided up into a number of sections that open to reveal the soft, cream-colored pulp enclosing large, oblong seeds. The Balinese also grill or fry the seeds and eat them.

Aside from its notorious odor, the fruit has the reputation of being an aphrodisiac. It is also said to be dangerous to eat durian when drinking any kind of alcoholic beverage. The leaves and roots are used in a compound for fever and the leaves in a medicinal bath for jaundice.

BALINESE: *duren*
INDONESIAN: *durian*
LATIN: *Durio zibethinus*; Family: *Bombaceae*

Guava

The Aztecs called it a sand plum, and the Spanish conquistadors carried it, along with the papaya, to Asia. There are a great many varieties of guavas: pear-shaped or round ones, small or large ones, yellow-seeded or red-seeded ones. Toward the end of the rainy season, in April, they are found literally everywhere in Bali.

Green when raw, the thin skin becomes yellow and soft in the mature fruit. The off-white flesh encases a multitude of seeds, which are also edible. Indeed some guavas seem to consist of seeds alone. Better varieties on the other hand have few seeds. Guavas have the texture of a slightly spongy melon, but are not very juicy. Ripe guavas have a distinct smell that can be rather over-powering.

The Balinese eat ripe guavas plain and immature ones are used for making *rujak* (see page 55). The raw fruit is also used to make a medicine for stomach ailments.

BALINESE: *nyambu sotong*
INDONESIAN: *jambu klutuk, jambu biji, jambu batu*
LATIN: *Psidium guajava*; Family: *Myrtaceae*

Jackfruit

Imported from India, where it was always reputed to be the food of sages and philosophers, the large, spreading jackfruit tree grows all over Bali in the lower elevations. It is often found in houseyards and provides excellent shade as well as delicious fruits.

The large fruits—which may grow to 50 kilograms—are covered with a bag when about half ripe. This encourages ants to swarm over the fruit and they guard it from pests.

BALINESE: *nangka*
INDONESIAN: *nangka*
LATIN: *Artocarpus heterophyllus* Lamk.; Family: *Moraceae*

The exterior is green and covered with small knobs. Inside is a fibrous, inedible central core surrounded by large, egg-shaped seeds. These are encased in bright-yellow pods that are slightly rubbery and are the edible portion of the fruit. The latex-like sap of the fruit is extremely sticky and makes removing the pods a rather messy job.

The mature fruit has a rather strong, almost pineapple-like odor, sometimes considered disagreeable. The taste, however, is sweet and quite delightful. Because of the immense size of the fruit, it is usually sold in pieces.

The immature fruit is also curried like a vegetable. The flesh of the ripe fruit is eaten raw as a snack or mixed with other fruits in *es campur*. It is sometimes eaten battered and deep-fried, like a banana. It is also churned up with ice and sugar and made into an iced fruit drink. The seeds are boiled and taste like chestnuts. The attractive, lemon-colored wood is used extensively in the carving industry.

Java Plum

This delicious, olive-like fruit that grows on high trees makes a challenging target for young boys with slingshots. Even taxi drivers make short work of the fruit in trees that grow in some of the hotel parking lots in Sanur.

BALINESE: *juwet*
INDONESIAN: *jamblang*
LATIN: *Eugenia cumini*; Family: *Myrtaceae*

The Java plum is very common in Bali, both in the south and in the foothills of the mountains. The tree bears fruit early in the rainy season, during the first weeks of November. The season does not last long—one week the markets and *warungs* will be overflowing with baskets of them and the next week there will be none. They do not keep or travel well and should be eaten soon after being picked.

The ripe Java plum is a very deep purple, almost black color, with a shiny, smooth and thin skin. It is easily crushed when ripe. The pulp is translucent white and quite juicy with a grape-like texture. There is a single oblong seed, coated with deep crimson red, which colors just about everything it touches. However, the color is easily washed off. The unripe fruit has a high tannin content and puckers the mouth. When ripe, the fruit is pleasantly sweet.

The juice of the fruit, infusions made from the leaves and bark, and the ground powder from the seeds are all used medicinally for stomach aches and diabetes. Powder from the seeds is especially used for ulcers.

Kaliasem

These cherry-like fruits grow in bunches on thick, green stems. In Bali, the trees are found in the medium elevations of the central part of the island and are seldom seen in the south. The fruits are very popular and are available in markets and foodstalls during the season, which begins in June and continues until the beginning of the rains.

The oval fruit is dark purple to brown in color, with a smooth skin. The pulp is more or less white, with one large, lima bean shaped seed which is not edible. The Balinese eat the entire fruit, skin and all, but the seeds must always be discarded. The flesh has a tart, but not particularly distinctive taste. It is often used to make *rujak* (see page 55 for recipe).

BALINESE: *kaliasem*
INDONESIAN: *gowok*
LATIN: *Eugenia polycephala;* Family: Myrtaceae

Kepundung

This is another one of those fruits that the Balinese simply love to snack on but which, for some reason, has never become popular with tourists. It is not available in hotels but during the dry season great bunches of them are sold in the markets, the stems tied up with bamboo twine.

The round fruit is yellowish-green, with tan splotches and a smooth skin. The skin is tough and encases two to four pods that are shaped like segments of a sphere. A papery white membrane with a slightly fibrous white core separates the segments. Each pod consists of soft and juicy pulp, similar to that of a grape. The flesh is white, translucent, and contains a seed.

The taste is a mixture of tart and sweet. The fruit is usually popped into the mouth whole and swallowed, seeds and all, and the skin is then discarded.

BALINESE: *kepundung*
INDONESIAN: *menteng*
LATIN: *Baceaurea racemosa*; Family: *Euphorbiaceae*

Langsat

Come to Bali in August and in every market you will see freshly-cut branches covered with tightly packed clusters of langsat. The trees can be found growing on the hillsides at elevations below 600 meters in northern Bali.

Langsat is native to the Malay peninsula, growing on a small tree about 10 to 20 meters high. The fruit is small, with a light, yellowish-tan skin that is mottled with brown specks and streaks. The skin surface is velvety with faint, lengthwise striations.

BALINESE: *ceroring*
INDONESIAN: *duku*
LATIN: *Lansium domesticum*; Family: *Meliaceae*

The rind is tough, but can be easily peeled off. The flesh is a translucent white, divided into three to five longitudinal segments of unequal size. Usually one segment is bigger than the others and contains a fully developed seed that adheres firmly to the flesh. The other segments may have cavities in them, but the seeds are small and vestigial.

The flesh separates easily from the interior of the skin and the segments separate easily from each other. The flesh is firm, somewhat like stiff gelatin, and has a pleasant aroma. It is not very juicy and has a distinctive sharp-sweet taste, tending to the sour side. The fruit is usually eaten by peeling off the skin with the fingers, throwing away the seed, and eating the flesh.

The fruit shells, when burned, give an aromatic smell and are used by the Javanese to drive away mosquitoes.

Lychee

The exiled Chinese poet Su Tung-po declared that the lychee fruit provided the only consolation to his eternal banishment. The first book ever written on fruit cultivation was devoted to its varieties, and in Bali they say a swarm of bats can strip a large tree laden with lychees in a night.

Lychees grow in clusters on a very tall, handsome, evergreen tree that is native to southern China. It thrives in Bali, and lovely big lychee trees line the roadsides between Mengwi and Bedugul.

The fruits turn a strawberry-red color when ripe and the skin is covered with small but rough, dot-like projections. It is easily removed, giving way to a white, translucent flesh enclosing one large seed. The flavor is just slightly acidic, tempered by sweetness and the Balinese are greatly enamored of its refreshing taste. The seeds are a source of oil and an ingredient in traditional medicines.

BALINESE: *leci*
INDONESIAN: *lici*
LATIN: *Nephelium litchi Sinensis*; Family: *Sapindaceae*

Mandarin Orange

This citrus has no equivalent in the West since it is neither a bona fide mandarin nor a tangerine. It has one thing in common with both those varieties, however:—the skin peels off easily.

BALINESE: *juuk semaga*
INDONESIAN: *jeruk garut, jeruk kaprok*
LATIN: *Citrus nobilis;* Family: *Rutaceae*

The tree is a native of China, where many varieties are cultivated. In Bali, *juuk semaga* was once a major cash crop in the northern coastal areas east of Singaraja, but in recent years these trees have been almost totally wiped out by an infestation of Citrus Plume Virus Degeneration or CPVD. The virus kills the trees, and so far no effective remedy has been found. The symptoms of the disease are yellowing leaves and a small, dry fruit. The high limestone plateau or Bukit area at the southern tip of Bali which is the second major *juuk semaga* area has recently been struck by the disease. There is reason to believe that the tasty *juuk semaga* may be a thing of the past in Bali before too long.

Even at its peak, the quality of *juuk bali* was somewhat variable. The surest test of a good one is to squeeze the fruit and see if it is firm, with the flesh filling the interior fully. But the most effective way to save both time and money is to try a single fruit before buying a kilogram or more. The oranges can be quite costly.

Mango

The mango is a native of Southeast Asia and is one of the earliest of all tropical fruits to have been cultivated by man. Westerners may eat Hawaiian or Mexican mangoes bought at their local supermarkets, but these diminutive and expensive fruits are a poor substitute for the plump, juicy mangoes found throughout tropical Asia.

The mango tree is very large, with a great, thick trunk and branches that bloom with tiny flowers beginning in September. The fruit is ready for market in November.

One explanation for the great variety of mango species is that the fruit has been cultivated for at least 4,000 years. Though it is difficult to generalize, the typical mango is a flattened ellipsoid with a large seed encased by soft, edible flesh. Unripe mangoes are firm and have a green skin, becoming soft and golden-yellow when ripe.

The variety known locally as *poh gedang,* the "papaya mango," is relatively small and has a bright yellow-orange skin and reddish flesh when ripe. *Poh mana lagi,* the aptly named "give-me-more mango," has a yellowish flesh and, as its name implies, can never be had enough of. Most locals consider the *poh golek* or "round mango" (it is also called *harum-manis,* meaning "fragrant-sweet") to be the tastiest variety grown here. You should not assume that all mangoes are equally delectable, however. Some have very stringy flesh and are quite sour.

Mangoes are for sale everywhere in Bali. They are fairly cheap and delicious—the quintessential tropical fruit. The Balinese love to snack on them. Raw mangoes make excellent *rujak* (see Appendix, page 55 for recipe). They are also an ingredient in *es campur,* a mixture of fruit, sugar syrup and shaved ice that is dispensed from push carts and food-stalls all over Indonesia. The pulp of the fruit, mixed with sugar and ice in an electric blender, also makes a fine iced juice. Mango juice is available bottled, canned and boxed, and many countries in Asia now have mango ice cream.

Pakel belongs to the same genus as the mango but is considered to be of much lower quality and is eaten almost exclusively as *rujak*. Its latin name, *foetida* (meaning 'fetid') indicates the attitude of the botanist who named it.

Pakel trees grow wild all over Bali and fruit at about the same time as the mango. The fruit is shaped like a mango, but is green with dark brown patches. The flesh of the *pakel* is quite fibrous and the ripe fruit tastes like a medium grade mango—rather sweet, but with a strange turpentine-like odor. The fruit cannot be recommended except to be eaten in *rujak*.

BALINESE: *poh/pakel*
INDONESIAN: *mangga/bacang*
LATIN: *Mangifera indica/Mangifera foetida*; Family: *Anacardiaceae*

Mangosteen

The mangosteen has been called the "Queen of Fruits." Indeed it was the favorite fruit of Queen Victoria on those rare occasions when she could come to Asia to enjoy it. That is one of its problems. A native of tropical Asia, the tree has resisted attempts to grow it out of its native habitat and the fruit, although encased in a thick shell, does not travel well.

BALINESE: *manggis*
INDONESIAN: *manggis*
LATIN: *Garcinia mangostana*;
Family: *Guttiferae/Garcinia*

The mangosteen grows under the dense foliage of a medium-size tree and is quite hard to see unless one is directly beneath it. One tree does not produce more than a few ripe fruits at a time and the tree is hard to propagate. There are thus no mangosteen plantations.

The dark-red husk encloses six or seven symmetrical segments. Splitting the rind is tricky because it is hard and tends to crumble, but once open, the segments separate easily. There is often considerable variation in the degree of maturity of each segment. The larger, mature segments usually have a seed while the others have an embryonic seed so small and soft it may go unnoticed. The taste is delicate, subtle and deliciously sweet.

Mangosteens are found in all the markets of Bali starting in September. Hotels serve them regularly, usually in a mixed fruit basket.

Melinjo

Melinjo is primarily valued for its seed, which is ground into flour, then pressed into a thin wafer and fried to form a crisp, flat chip called *emping*. It has a nutty and faintly bitter taste that is an excellent accompaniment to beer. Most Balinese restaurants keep it in large jars.

The *melinjo* tree is of medium size and grows wild in the forest. You are not likely to see one unless you look for it. The small, oval fruit is both green and red. The skin is firm but it peels away easily from the seed within.

Melinjo is also eaten boiled—the skin is peeled off, and then the shell of the seed is cracked with the teeth and the nut is eaten as one would eat a peanut.

BALINESE: *melinjo, meninjo*
INDONESIAN: *melinjau, belinjau*
LATIN: *Gnetum gnemon*; Family: *Gnetaceae*

Orange

This variety of citrus is the closest one comes to finding an orange in Bali. In spite of the epithet *manis* ("sweet") often applied to them by the Balinese, the smaller varieties are quite sour. This orange is used almost exclusively for juice. But be aware that when you order orange juice other than in a hotel, you will almost invariably find it watery, sweet and hot—all of which can be a bit disconcerting if your mouth is watering for a nice glass of fresh, cold, pulpy orange juice. But this is the way many Balinese like their OJ and *de gustibus non disputandum est.*

BALINESE: *juuk peres*
INDONESIAN: *jeruk manis*
LATIN: *Citrus aurantium*; Family: *Rutaceae*

Otaheite Apple

Some think it does not merit cultivation, others liken it to an inferior mango. This large tree is a native of Polynesia—Otaheite is the old name for Tahiti—and it grows everywhere in Bali. The fruit is sometimes found in the markets, but is usually picked from the tree while still unripe and used in *rujak* (see Appendix, page 55 for recipe). The ripe fruit can be eaten, but is not very tasty.

The oval fruit is green in color and is usually slightly striped. It has a relatively thick rind and the flesh is hard and crisp with a tangy taste when still unripe.

When the fruit is sliced laterally, a star-shaped, pentagonal symmetry is seen, with a slightly darker central core. Five small, pale green seeds surround the core and are firmly embedded in it.

BALINESE: *kedongdong*
INDONESIAN: *kedongdong*
LATIN: *Spondias dulcis, S. cytherea*; Family: *Anacardiaceae*

Otaheite Chestnut

This rather tasteless fruit is commonly shelled and boiled and sold in the market in a container of water. It looks very much like the cubed, cooked, white meat of a chicken, and has a rubbery feel to it, but is rather crumbly—like hard cheese. It is very starchy and cannot be recommended enthusiastically.

BALINESE: *gatep*
INDONESIAN: *gayam, tolok*
LATIN: *Inocarpus edulis*; Family: *Leguminosa*

The otaheite chestnut is a native of the Malay archipelago and the Pacific, the English name coming from the old name for Tahiti. It is fairly common in Bali, but it is unlikely that you will see it sold as a fruit; it is usually sold in a cooked form.

The fruit consists of a single-seed pod. When seen from its thin side, it resembles a green teardrop. It has a thin and prominent fin that extends from the top to the bottom.

The surface of the fruit is very wrinkled and there is a distinct longitudinal pattern to these ridges. The skin is leathery and tough, like that of a lemon.

A coconut-like husk grows beneath the skin, which can be pried open with a knife. Inside is a large cavity containing a white, crisp nut covered with a thin, brown, papery skin. The interior nut separates easily from the husk and is roughly triangular. The white surface is smooth on one side and pitted on the other. The seed is not at all juicy.

Otaheite Gooseberry

Beginning in late September, if you look very carefully at certain trees in Bali, you will see small, pale-green fruits hanging in grape-like clusters. These are otaheite gooseberries, which are seldom seen in the market because they are picked by villagers and made immediately into *rujak* (see Appendix, page 55 for recipe).

The small, round fruit is firm, crisp and juicy when ripe. It has a single hard seed which adheres tenaciously to the flesh. The fruit has a mildly acidic taste which to some tongues is zesty and refreshing, to others sour. But again, it is usually never eaten alone, but is made into *rujak.*

BALINESE: *cereme*
INDONESIAN: *ceremai*
LATIN: *Phyllanthus acidus* Skeels; Family: *Euphorbiaceae*

Papaya

Also papaw

The palm-like papaya plant grows in every village in Bali and the fruit is sold everywhere. It grows year-round and fruits continuously during a three- to four-year lifespan.

Balinese papayas come in many shapes and sizes, but have the general shape of elongated melons and are large, fleshy and hollow. The hollow is normally filled with hundreds of small, black seeds—although there are seedless varieties. The flesh varies from yellow to a deep red and the latter are in great demand. The skin is green, changing to golden-orange when the fruit is ripe. The fruit ripens quickly and is usually picked green and then brought to market.

BALINESE: *gedang*
INDONESIAN: *papaya*
LATIN: *Carica papaya*; Family: *Caricaceae*

Unripe papayas are often boiled and eaten as a vegetable and make a favorite ingredient for *rujak* salad (see Appendix, page 55).

Usually, only the female plants bear fruit. When a male plant proves fertile, the fruits hang down in a long chain. This is regarded as a freak of nature and the chain is used in some Balinese ceremonies designed to drive away evil spirits.

The fruit, and especially the latex of the tree and the leaves, contains papain, an enzyme used commercially as a meat tenderizer and as an aid for digestion. The Balinese commonly wrap a chicken or tough piece of meat in papaya leaves and let it remain overnight to tenderize it. The juice is sometimes used to relieve insect bites, and as a cosmetic to remove blemishes from the skin and to take the sting out of burns.

Passion Fruit

Also giant granadilla

The English name for this fruit comes from Spanish missionaries who thought its flower was a symbolic representation of the crucifixion of Christ.

Passion fruit is not very popular among the Balinese and is only available in limited quantities in some of the larger markets at the end of the dry season. You are not likely to see the vines themselves unless you travel off the beaten path to the Candi Kuning area, or to Kintamani.

The fruit is the size of a baseball and is orange or yellow when ripe, often tinged with green. The pliable shell is smooth and hard with a spongy inner lining and contains a multitude of seeds, each of which is encased in a transparent, gelatinous pulp. The seeds are what is eaten and can be more or less drunk from a ripe fruit once the shell is pried open. Passion fruit is also prepared as a vegetable.

BALINESE: *markisah*
INDONESIAN: *markisa*
LATIN: *Passiflora quadrangularis*; Family: *Passifloraceae*

Pineapple

The pineapple plant is a native of the New World, and like so many others was brought to Asia by the Portuguese and the Spanish during the 16th century. There are a great many varieties, but the pineapples grown in Indonesia tend to be smaller and sweeter than the typical Hawaiian pineapples found in a western supermarket.

Most of the pineapples sold in Bali come from Java. It is interesting to note that in many areas of the archipelago, pineapple juice was traditionally regarded as a diuretic and was used to induce abortions. The strong fibers of the pineapple plant were also traditionally woven into cloth on many islands before the advent of cotton.

Pineapples are found in all Balinese markets throughout the year, and are very inexpensive and very sweet.

BALINESE: *nanas*
INDONESIAN: *nanas*
LATIN: *Ananas comosus*; Family: *Bromeliaceae*

Pomegranate

Paris gave it to Venus, and the Israelites, wandering in the wilderness, longed for the pomegranate—but in Bali, this exotic fruit is not very popular and is used mainly in medicinal preparations.

The pomegranate is a native of the Middle East but was cultivated in India and East Asia already in ancient times. It grows on a large shrub or small tree which has brilliant orange-reddish flowers. About the size of an orange, the dull-red pomegranate has a tough, leathery skin which allows the fruit to travel well. The skin encases six paper-thin septums, each containing seeds that are individually encased within a transparent, pulpy capsule. The local varieties are not very fleshy and the Balinese like to distinguish between the white and the more common red-pulp pomegranates.

BALINESE: *delima*
INDONESIAN: *delima*
LATIN: *Punica granatum* K.; Family: *Punicaceae*

The fruit and the rind are an effective anti-bacterial agent, and the dried rind is used as a relief for dysentary. Mohammed even believed it to be endowed with spiritual properties: "Eat the pomegranate—for it purges the system of envy." In traditional Malay tales, the mouth of a beautiful woman is often likened to a ripe pomegranate split in two.

Pomelo

Also shaddock

The entire citrus family is of Asian origin, and its cultivation has been carried on since the beginning of recorded history. The Indonesian word *jeruk* is applied to all citrus and the Balinese equivalent is *juuk, joak* or *juwuk.*

When the grapefruit-like pomelo reached Europe in the twelfth century, it was known as "Adam's apple." The English name is a variant of the French *pomme,* or apple; its other English name, shaddock, derives from the surname of the sea captain who first brought the fruit to America.

BALINESE: *juuk bali*
INDONESIAN: *jeruk besar*
LATIN: *Citrus grandis;* Family: *Rutaceae*

The pomelo is a native of the Southeast Asian mainland, and the small tree, with fruits the size of a child's head, is seen almost everywhere in Bali. The fruit is on sale in most markets and many roadside stalls when in season.

The round fruit has pale-green to yellow skin finely dotted with oil glands. The interior pulp—in Bali there are two varieties, green and red—is covered with a spongy, white rind. The segments are covered with a membrane which must be removed when the fruit is to be eaten, as it is very bitter.

The pomelo is not as juicy as a grapefruit, but has a tart and refreshing flavor. It travels well and makes a good picnic food. The green variety has more flesh and can be a bit smaller than the pink variety. There are some varieties of each color in which the white rind is comparatively thinner, making for less work and more pulp.

Rambutan

The name *rambutan,* deriving from the Malay word for hair, *rambut,* reflects this unusual fruit's most distinctive feature—a coat of thin, flexible, almost plastic-like cilia that cover the bright-red or yellow-green rind. The Balinese name, *buluan,* has a similar etymology—being derived from *bulu,* or body hair.

Rambutans grow in the medium and higher elevations of Bali, and can be seen in the Denpasar-Kuta-Sanur areas. The Mengwi-Bedugul highway is lined with rambutan trees, and in season, beginning in September, the road is littered with rambutan skins left by motorists.

BALINESE: *buluan, aceh*
INDONESIAN: *rambutan*
LATIN: *Nephelium lappaceum;* Family: *Sapindaceae*

The fruits hang in bunches and come in two colors: deep crimson, the most common, and bright yellow. The thick, tough skin is flexible and can be split and peeled off quite easily with a thumb. The translucent pulp is firm and white. It is not very juicy and the flavor is tart and refreshing. Rambutans have a high content of ascorbic acid and contain the same amount of sugar as the lychee, which they resemble in appearance and taste.

Each fruit has one large seed which serves as the basis of distinction between two varieties. The seed of *aceh* can be easily removed from the pulp, while that of *buluan* cannot. The existence of but one seed in this fruit has given rise to a popular Balinese proverb. When asked if he is thinking about taking a second wife, a man may answer: "*Buka batun buluane.*" Translation: "One is enough."

Salak

In the medium elevations of Bali there are numerous plantations of *salak*, and the fruits are in great demand. The palm looks rather formidable, and is nearly stemless, low, and full of very sharp thorns. The bunches of fruit nestle in the middle, almost defying someone to pick them.

The fruit is brown, tear-drop shaped, and covered with triangular scales. The points are not really sharp and rubbing the surface against the grain is like rubbing a coarse file. In overall shape, and in the appearance and texture of the fruit (though not in taste), the *salak* is like a huge garlic.

The skin is tough but very thin, and the fruit peels easily, revealing a pear-shaped interior that is a light tannish-yellow color. The interior is divided into three or more lobes, one or two of which are much larger than the others.

BALINESE: *salak*
INDONESIAN: *salak*
LATIN: *Salacca edulis*; Family: *Palmae*

Raw *salaks* are full of tannin and very astringent. When ripe, salak has a distinctive taste, not sweet, but agreeable. It is crisp and breaks easily, having the consistency of a turnip or raw carrot. It is not juicy and *salaks* are often served whole at hotels because they can be peeled and eaten with ease. They are available during the dry season and when out of season are imported from Java.

The large seeds are often made into toys by Balinese children. They drill two holes through the seed and pass a string through them. The seed buzzes when the ends of the strings are pulled and the seeds rotate.

Sapodilla

Also naseberry

When the Spanish encountered the Aztecs the latter had been chewing gum, *chicle*, for years. *Chicle* comes from the latex of a tree that bears tasty brown fruit. In Mexico the tree is still used for its gum, the principle ingredient of chewing gum; but in Asia the tree is only used for its delicious fruits. The Malay name, *ciku*, reflects this origin.

Sapodilla grows on a small, unpretentious tree found all over Bali. Its fruit is among the most popular in Bali, and is eaten as a snack or with meals. They are not commonly found in tourist spots because they are so soft they cannot be easily sliced and sold.

BALINESE: *sabo, sawo*
INDONESIAN: *sawo manila*
LATIN: *Achras zapota/ Sapota zaspotilla*
Family: *Sapotaceae*

The fruit looks like a small, brown potato with smooth skin. The flesh is a rich brown color with a vague radial structure of a lighter color, and contains one to five large seeds. The sapodilla has a very sweet, sugary taste, which hints at maple sugar. It is soft, though not juicy. The texture is very slightly gritty, but not objectionably so. The poetic French botanist Descourtiez says it has the "sweet perfumes of honey, jasmine, and lily of the valley."

The fruit can be broken open and the flesh easily eaten without consuming either the seeds or the skin. Sapodilla must be eaten ripe, however, because the fruit contains tannin and a milky latex when unripe.

Sawokecik

Unlike most fruits trees the *sawokecik* is principally grown for its lovely, dark, durable wood, which, along with other similar woods, is sold as "rosewood." *Sawokecik* is used for woodcarving and the medium-size tree is common in West Bali and on Nusa Penida.

The trees bear fruit in August. The season does not last long, but, while it does last, the colorful fruits—maroon, red to orange, often striped—are seen in many local markets. They are ellipsoid in shape and have a small, five pointed cap at one end.

The skin is very thin and soft and contains the creamy-white flesh which has a slightly sugary taste. There is a vague grittiness to its texture, and it tends to crumble when one attempts to slice it. The interior of the fruit is divided into five longitudinal compartments, each of which may have one large seed.

BALINESE: *sawokecik, sawo bali*
INDONESIAN: sawo
LATIN: *Manilkara kauki*; Family: *Sapotaceae*

Soursop

Also zurzak

This fruit's common names come from the Dutch *zuur zak*, meaning "sour sack." The English variant "sop" is used in the sense of something that soaks up a liquid. A native of tropical America, the trees are scattered all over Bali, not on plantations, but along the roads and in houseyards.

The fruit varies in size and shape but is roughly pear-shaped. The skin is thin and soft, and covered with conical nubs. The interior is soft, white, and juicy and is full of seeds and membranes distributed around a pithy core.

The flavor is pleasantly zesty, slightly acidic, and sweet—though the stringy nature of the interior detracts somewhat. It is also made into a popular juice. Soursop is available in Balinese markets at the end of the dry season.

BALINESE: *sekaya cina, sakaya, srikaya*
INDONESIAN: *sirsak*
LATIN: *Annona muricata*; Family: *Annonaceae*

Starfruit

Also carambola

Once known in the West only by travellers to Asia, starfruit is now becoming quite fashionable in the Western hemisphere, going under the Spanish name of carambola. The fruit's English name reflects the longitudinal, sharp-edged ridges which give the fruit, in cross-section, the shape of a five-pointed star.

Starfruit grows abundantly on a small tree that is found just about everywhere in the low and medium elevations of Bali. The fruit has not been highly bred so there are a number of local varieties, differing in size and sweetness. Only a fraction of the fruit finds its way to the local markets, the majority being eaten on the spot, usually in the form of *rujak* (for a description of *rujak*, and how to prepare it, see page 55).

The translucent skin of the golden-yellow fruit is so thin it can be easily punctured by a fingernail. The crisp and juicy pulp is fragrant and has a tart taste. The fruit is firm when ripe and can be eaten raw—skin and all—once the tough edges of the five ridges are peeled off. Despite the slightly acidic taste, starfruit does not contain tannin and so is not astringent.

The tree and the fruit are considered to have uses varying from removing cloth stains to curing hangovers. Since it contains potassium oxalate, starfruit juice is used to polish the blades of the Balinese *keris*, or ceremonial daggers. It is also very high in vitamin C.

BALINESE: *belimbing*
INDONESIAN: *belimbing manis*
LATIN: *Averrhoa carambola*; Family: *Oxalidaceae*

Sugar Palm

This tree, which grows in the medium and higher elevations of Bali, is not valued as much for its fruit as it is for its sap—which is made into *tuak,* a mildly alcoholic drink.

The round fruits of the sugar palm hang in long, closely-packed clusters. A tough green skin covers a fibrous layer which contains three white pods. The fruit in these pods is prepared for the market by shelling and boiling. Great care must be used when extracting the pods because the juices of the raw fruit produce a sharp stinging sensation if they contact the skin. The pain subsides in about 15 minutes and the fruit itself is harmless after it has been cooked.

BALINESE: *beluluk*
INDONESIAN: *aren*
LATIN: *Arenga pinnata*; Family: *Palmae*

In the markets, sugar palm fruits are always sold from a container full of water, and the small, white, onion-like fruits are scooped into a plastic bag along with some water. Every *warung* has a jar of them ready to be made into *es campur,* a mixture of sliced fruits served with ice and a sweet syrup. By itself, the fruit is rather tasteless, but in *es campur* it makes a refreshing snack.

The tree is very useful: It is tapped for its sap which is boiled down to a ubiquitous brown sugar—*gula*—or is fermented into *tuak*. The strong black fiber at the base of the flower stalk is used as twine in Bali. It is particularly important in making black fiber roofs for various kinds of shrines and other buildings.

Water Apple

This blush-colored fruit is crisp, juicy and sweet. The water apple grows on small to medium trees that are rather crooked. They line many streets and are common in house yards all over Bali; when in season, the fruits litter the ground. In the dry season the markets have several varieties of this inexpensive fruit for sale.

It is round or oval in shape and is crowned at the apex with a calyx segment which is often swarming with ants. There is a single round seed in the large seed cavity.

The fruit has almost no taste, except, perhaps a faint sweetness. It is juicy and used as a thirst quencher; indeed, the Balinese call it *nyambu air*, the last word meaning "water." The unripe fruit is used for making *rujak* (to make *rujak* see page 55).

BALINESE: *nyambu*
INDONESIAN: *jambu air, jambu semarang, jambu bol*
LATIN: *Eugenia jambos*; Family: *Myrtaceae*

Watermelon

The Balinese watermelon differs in no significant way from those grown in most parts of the world. It has been cultivated since prehistoric times and is thought to be a native of Africa.

In Bali, the fruit comes in two shapes—one is almost spherical and is usually dark green with bright-red, juicy, crisp flesh and the other variety is more cylindrical, with rounded ends. It is usually a lighter green color with dark green longitudinal stripes and the interior is the same as that of the spherical variety. Bali now offers a seedless variety imported from Java.

Watermelons are just as much a favorite on Bali as they are everywhere else, and the markets are crowded with them during the dry season, which is when they grow best. They are commonly sold along the roadside right next to the fields where their vines grow.

BALINESE: *semangka*
INDONESIAN: *semangka*
LATIN: *Citrullus vulgaris*; Family: *Cucurbitaceae*

Appendix

RECIPES FOR *RUJAK* AND *PETIS*

Two favorite Balinese snacks are *rujak* and *petis*, both of which are made with a spicy sauce and whatever tart, raw fruit is available. These dishes are made and sold on the spot in practically every market in Bali. They can also be purchased ready-made in some stores and restaurants and needless to say, every housewife has her favorite recipe. *Rujak* is craved by pregnant women, apparently fulfilling the role played by pickles in the West.

To make *rujak* take a few grams (one small cube) of *sera*, "shrimp paste" (also known as *terasi*) and heat it over a medium flame. When it is cooked, put it in a mortar stone (or blender) and add the following:

a large pinch of salt (*uyah*)

4 ounces of brown palm sugar (*gula barak*)

1 tablespoon of vinegar (*cuka*)

1 small chili pepper (*tabia*)

1 tablespoon of tamarind pulp (*lunak*)

Mix the above ingredients into a sauce and pour it over the cut pieces of a raw fruit of your choice. The result is a delicious sweet-and-sour snack.

Petis is simpler to make. All one needs is cooked *sera* "shrimp paste," one chili pepper, and *kuah pindang*, the liquid left over when fish is boiled in salt water. Mix the two solids into the fish water and add the sliced fruit.

The taste of these snacks is richer, and more pungent than we are used to in the West, so don't approach them thinking they are anything like our fruit salad. Just enjoy the difference.

Overleaf: A rujak *street vendor in Kuta.*

Index by Balinese name

Index by Latin name

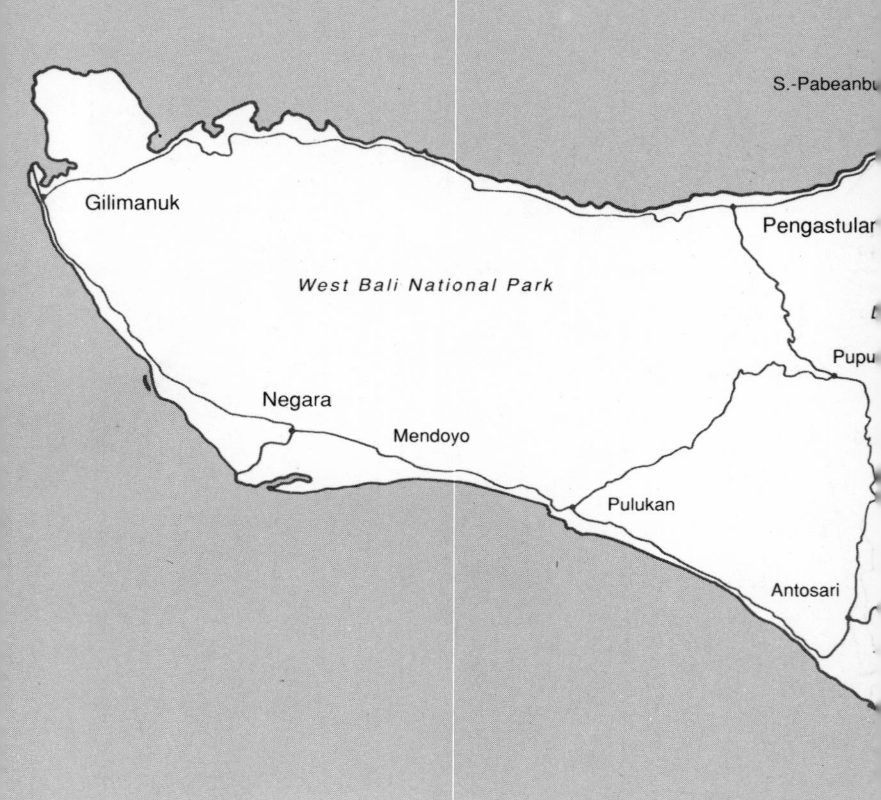
S.-Pabeanbu
Gilimanuk
Pengastular
West Bali National Park
Pupu
Negara
Mendoyo
Pulukan
Antosari
N

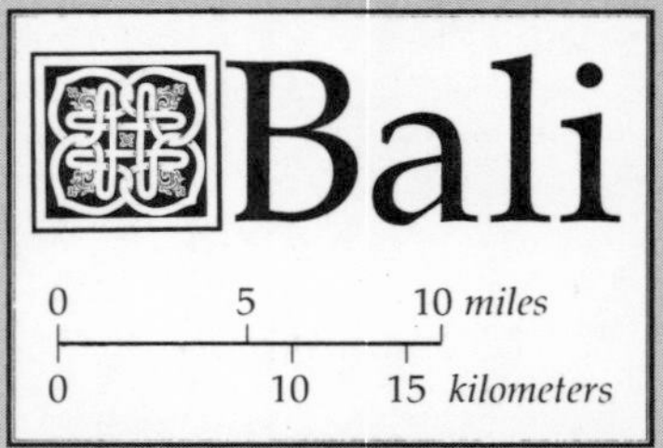
Bali
0
5
10 miles
0
10
15 kilometers